YOUR KNOWLEDGE HAS VALUE

- We will publish your bachelor's and master's thesis, essays and papers

- Your own eBook and book - sold worldwide in all relevant shops

- Earn money with each sale

Upload your text at www.GRIN.com and publish for free

On the quantisation of the frequency of electromagnetic radiation

William Fidler

Bibliographic information published by the German National Library:

The German National Library lists this publication in the National Bibliography; detailed bibliographic data are available on the Internet at http://dnb.dnb.de.

ISBN: 9783668910560
This book is also available as an ebook.

Print and binding: Books on Demand GmbH, Norderstedt, Germany
Printed on acid-free paper from responsible sources.

The present work has been carefully prepared. Nevertheless, authors and publishers do not incur liability for the correctness of information, notes, links and advice as well as any printing errors.

GRIN web shop: https://www.grin.com/document/463095

On the quantisation of the frequency of electromagnetic radiation

W M Fidler

Abstract

It is posited that the frequency of electromagnetic radiation may be quantised and two methods are derived which permit the calculation of the magnitude, Δ, of the quantisation.

The work does not establish the existence of frequency quantisation and the two methods derived may be described by the oxymoron, systematically arbitrary.

Both methods rely upon the Fidler diagram [3].

The first method employs the superposition of a fractal path on the diagram and shows that the magnitude Δ is given by: $\Delta = \dfrac{(V)}{(\sqrt{2})^{n-1}}$, where (V) is the Planck frequency and n is an odd, disposable integer.

The second method involves the tiling of the Fidler diagram by a succession of self-similar tiles which are progressively-reducing versions of the diagram itself. For this case, the magnitude of the frequency-quantisation is given by: $\Delta = \dfrac{(V)}{(\sqrt{2})^{2(n-1)}}$, where n is any disposable integer.

It is shown that the Fidler diagram is the two-dimensional projection of a three-dimensional surface and that the specific energy, specific energy intensity and specific energy density of a photon may, by the procedure outlined at the end of the work, be calculated from the diagram without additions thereto.

List of Contents

Page No.

Introduction..4

Quantisation of the frequency of electromagnetic radiation..6

A frequency-quantised Fidler diagram 7

The Fidler diagram as the two-dimensional projection of a three-dimensional surface..........13

The extended Fidler diagram 15

References...17

<u>**Introduction**</u>

It is considered axiomatic that the spectrum of electromagnetic radiation, and, in particular, the frequency, is continuous. As is well known, continuity may be considered as a manifestation of the macroscopic viewing of processes which are discontinuous on the microscopic scale; an example of this is the macroscopic treatment of a gas in classical thermodynamics. Here, the gas is considered to be a continuous substance and its behaviour is described by few parameters. We hence propose that the notion of continuity be placed in the same conceptual class as Platonic solids. On a philosophical basis, it may be implied that any phenomenon which is analysed using differential calculus admits of discontinuity, in that differential quantities are never zero; ordering of combinations of such quantities is done on the basis of the mathematical phrase 'tends towards zero'.

The Planck-Einstein-Schrödinger equation is $E = h\nu$, where the symbols have their usual connotation. Now, one of the quantum rules stipulates that the quantity $h\nu$, called the photon, may only be emitted or absorbed in integer multiples. The so-called electromagnetic spectrum is the range of all possible frequencies of this photon, and, given that the magnitude of Planck's constant of action, h, is $6.62606876(52) * 10^{-34}\,Js$, it is probably true to say that the properties of a single photon have never been measured to sufficient accuracy to resolve the question of whether or not the frequency is quantised.

At the very low frequencies (i.e. a few Hertz), characteristic of the geophysical phenomenon called Schumann waves, a measurable signal could only be obtained from the combined effect of literally trillions of trillions of photons and hence quantisation of the frequency might be masked by the nature of the sheer number of photons, for there is no reason why they should all act in concert. Even the charge-coupled devices associated with that branch of Astronomy which investigates exceptionally-long wavelength radiation, whilst registering the presence of such, are probably incapable of resolving the question of frequency-quantisation. At the other end of the spectrum where measurements may be made of the properties of a relatively small (relatively speaking) number of gamma ray photons, it is doubtful that the frequency could be determined with sufficient accuracy to establish, or otherwise, frequency quantisation, given that the frequency of such photons is of the order of 10^{23}Hz. Given that quantisation of frequency, even at lower frequencies has never been reported it is concluded that if the phenomenon does exist, then it is very small throughout the range of frequency of the spectrum of electromagnetic radiation.

It is known that electronic circuitry incorporating the Josephson effect has been produced that can measure voltages of the order of a few picovolts and it is suggested that this may provide instrumentation which will facilitate the investigation of the posited quantisation of frequency.

If the reality of frequency quantisation can be established then this will extend the quantum rule discussed earlier.

Further, ever since the redefinition in 1967 (together with later amendments), the second has, in a sense, become a quantum-mechanical quantity for it is defined, [1] as the duration of 9,192,631,770 periods of the radiation corresponding to the transition between the two hyperfine levels of the ground state of the Caesium-133 atom, and hence, frequency through association with the second may be argued to possess aspects of a quantum-mechanical character. Indeed, it follows that this notion may apply to any quantity so associated.

It is in the spirit of rejection of continuity that the following work examines the quantisation of the frequency of electromagnetic radiation.

We invoke the Jesuit credo,' it is more blessed to ask for forgiveness than permission', as articulated by Wilczek [2].

<u>**Quantisation of the frequency of electromagnetic radiation**</u>

We posit that the frequency, ν, of electromagnetic radiation is given by the simple linear equation:

$$\gamma\Delta \ = \ \nu \ \text{------------------------} \ (1),$$

Where, γ is a pure number and Δ, which must have units of Hertz, is a non-zero increment of frequency. We call γ the frequency number. Further, it is posited that for the purpose of this work, Δ, is a constant, has the character of a finite difference, and hence, is very small.

If the spectrum of electromagnetic radiation is continuous then all γ are members of the set of real numbers. Conversely, if the frequency is quantised then all γ are members of the set of natural numbers.

 In the case of a continuous spectrum of radiation there is, at bottom, no utility in equation (1) if used in other than the manner prescribed, for the combination of free choices for Δ in conjunction with any number in the range of real numbers renders the equation hyperbolic, and an infinite set of rectangular hyperbolae, with frequency ν as parameter may be generated, which, in the context of the present work, is devoid of any physical meaning.

In the case of continuous electromagnetic radiation the abscissa, ordinate and hypotenuse of the Fidler diagram [3] are continuous lines and the space enclosed by these lines is filled completely. The magnitude of the division of the bounding lines into discrete intervals is entirely arbitrary and is hence not unique; any such discretisation is made on the basis of convenience and is done by fiat. Further, unlimited interpolation within any interval of discretisation implies continuity.

In keeping with the notation in [3], Planck 'quantities' are denoted by upper case letters enclosed in round brackets. Further, we call the point in the Fidler diagram with coordinates (1,1) , the Planck point.

A frequency-quantised Fidler diagram

A discretised, or, 'discontinuous ' Fidler diagram is shown in Fig1.

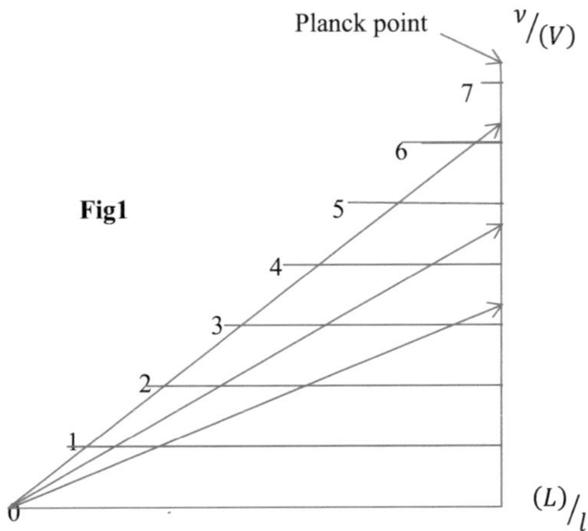

The sloping lines are lines of constant radiation Strouhal number, S_r, [2], and represent the spectrum of electromagnetic radiation in different substances. In this representation each spectrum only has meaning at the intersection with a horizontal line; which is isotonic and is the locus of an infinite number of intersections. In order to emphasise this we have omitted the 'luminal' line which, in the case of a continuous spectrum in vacuo would pass through the numbered points (and all intermediate points).

The scale of the diagram shown here seems to imply that the very low values of the radiation Strouhal number (and hence very high values of the index of refraction) reported by Hau et al [4], [5] and Schmidt et al [6], could not be accommodated on the diagram. However, the figure depicted is solely for the purposes of illustration and, as will be shown later, can be rendered sufficiently fine-grained to encompass virtually all the above-mentioned experimental data.

We now proceed to devise methods, by means of which the magnitude of the quantisation of frequency may be established. At the outset it must be made clear that the results of these procedures do not establish the existence of frequency quantisation and further, the means by which the results are obtained may be described by the oxymoron 'systematically arbitrary'.

As shown in Fig2 we superimpose a fractal path on the Fidler diagram.

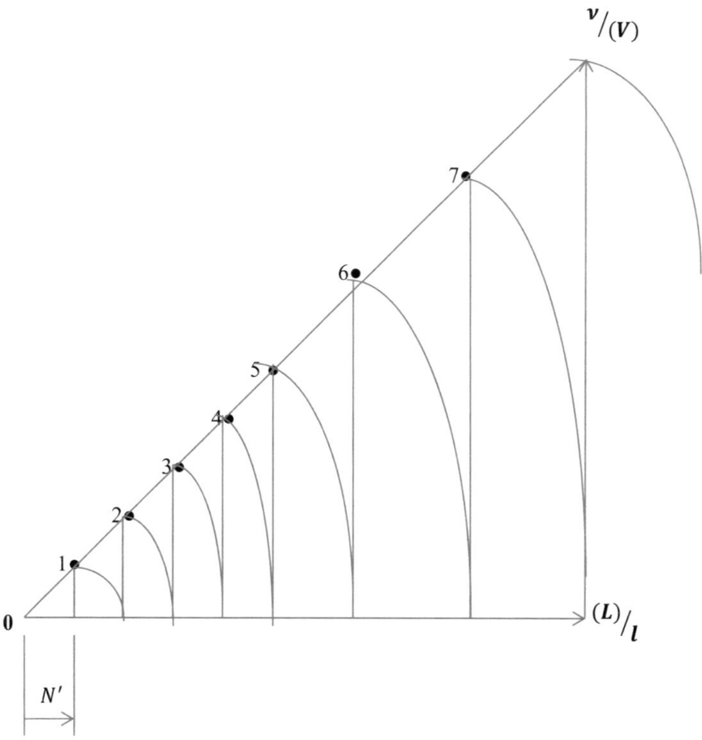

Fig2

The path shown, which is continuous and consists of a series of circular arcs centred on **0** followed by lines of constant wavelength may be extended indefinitely in either direction.

If we designate the numbers at the intersection of the path with the hypotenuse as addresses, then, by simple geometry it is easy to see that these points, marked '•' are located along the luminal line at distances $(\sqrt{2})^{a} * N'$ from **0**, where **a** is the address number. If there are **n** addresses to the Planck point then we may write:

$$(\sqrt{2})^{n} * N' = \sqrt{2}. \quad \text{Hence } (\sqrt{2})^{n-1} * N' = 1 \text{ -------------------- (2).}$$

If we choose to discretise the abscissa in increments of, N', then, because of the nature of the diagram, this will automatically discretise the ordinate by the same amount.

Now, along the ordinate, $N' = \Delta/(V)$; hence, from equation (2) we now have an expression for the discretisation of the frequency, viz: $\quad \Delta = \dfrac{(V)}{(\sqrt{2})^{n-1}}$ ---------------------- (3).

Further, the division of the coordinate axes must be by an integer amount, and so, **n**, must, in the above, be an odd integer.

As noted previously, we can change, **n** , systematically, but termination thereof is arbitrary.

There are some interesting consequences of the superposition of this path on the Fidler diagram.

For the purpose of illustration consider the construction shown in Fig3. A circular arc, centred on **0** intersects the abscissa at a distance, $\gamma_b N'$ from **0**. The source of the arc is located on and along the luminal line at a distance, $\sqrt{2}\gamma_a N'$ from **0** .

Now, $\sqrt{2}\,\gamma_a N' = \gamma_b N' \rightarrow \sqrt{2}\,\gamma_a = \gamma_b$. If the source of the arc is located at a quantised frequency point then γ_b cannot be an integer for γ_a is an integer and the square root of two cannot be represented by the ratio of two integers. It then follows that γ_b has the form $\gamma + \alpha$.

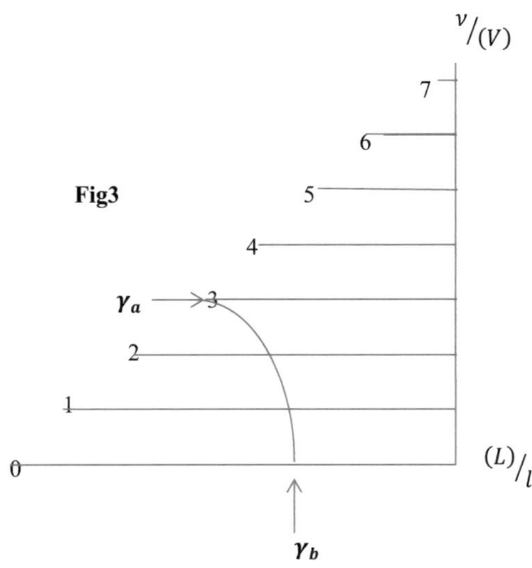

We can amplify the previous statement by an extreme, but interesting example.

Irrational numbers have been, and continue to be studied extensively. In particular, the square root of two holds an abiding fascination for attempts to calculate it to as many decimal places as possible (two million places of decimals have been reported---without termination). We make no attempt to follow such an extent but, we employ a value for $\sqrt{2}$ quoted in Wikipaedia to 65 places of decimals. viz:

1.41421 35625 73095 04880 16887 24209 69807 85696 71875 37694 80731 76679 73799

Consider a photon having a frequency in vacuo of $\mathbf{10^{10}\ Hz}$.

If, for nothing other than convenience of illustration we assume the magnitude of the quantisation of the frequency to be one Hertz, then $\gamma_a = \mathbf{10^{10}}$; we have noted previously that the magnitude of γ_b is of the form $\gamma + \alpha$, hence, the magnitude of γ_b must be:

141421 35625. 73095 04880 16887 24209 69807 85696 71875 37694 80731 76679 73799.

Thus, we may determine the magnitudes of γ and α to quite extraordinary (and, in all probability, wholly unnecessary) precision.

For brevity we write this as $\mathbf{N + D}$. If we project this 'distance' vertically upwards, then, for a frequency quantisation of one Hertz the terminating point of this projection must be at

$\gamma = \mathbf{14142135625}$, i.e. $\mathbf{N}$. for there is no point on the ordinate between $\mathbf{N}$ and $\mathbf{N+1}$.

Hence, the radiation Strouhal number, [3] and the index of refraction at this point are given by: $\mathbf{N}/\mathbf{N + D}$, and, $\mathbf{N + D}/\mathbf{N}$, respectively.

It may be noted that the method for division of the coordinates devised previously is not unique, and it is shown in Fig4 that the Fidler diagram may be 'tiled' by progressively-smaller versions of the diagram itself. The internal lines shown, may, in keeping with the previous example, be considered as a path.

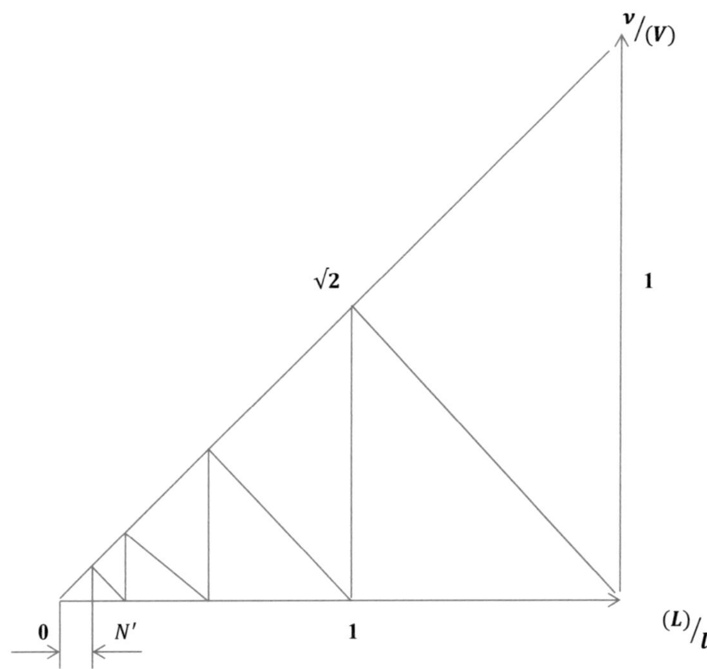

Fig4

Starting from **0** and proceeding along the luminal line it is easy to show that the distance from **0** of the first intersection of the path with the luminal line is $\sqrt{2}N'$, whilst the distance of the second intersection is $\sqrt{2}\sqrt{2}\sqrt{2}N'$, and the third, $\sqrt{2}\sqrt{2}\sqrt{2}\sqrt{2}\sqrt{2}N'$, etc.

Hence, for **n** intersections to the Planck point we may write: $\left(\sqrt{2}\right)^{2n-1}N' = \sqrt{2}$, where **n** is a natural number.

This reduces to: $\left(\sqrt{2}\right)^{2(n-1)} N' = 1$ ------------------ (4).

Using exactly the same arguments as before, we may write:

$$\Delta = \frac{(V)}{\left(\sqrt{2}\right)^{2(n-1)}} \text{------------- (5).}$$

This expression yields a valid result for the frequency discretisation irrespective of the nature of the integer, n; this is in contradistinction to that given by equation (3) which is only valid for odd integers.

Hence, we have established two ways in which a measure of the quantisation of frequency may be determined. In the first method we posit the number of intersections of a fractal path with the hypotenuse of the Fidler diagram, whilst in the second we posit the number of intersections with the hypotenuse of a path formed from a set of self-similar tiles.

Further, for the same value of, **n** , the discretisation calculated by equation (5) is smaller than that calculated by equation (3) by a factor of $\left(\sqrt{2}\right)^{n-1}$.

Since neither equations (3) or (5) establish the existence of quantisation of the frequency of electromagnetic radiation then their use in any subsequent investigation is simply a matter of preference-----although, the second approach yields an expression by which the magnitude of the quantisation may become very small in comparison to that determined from equation (3).

All of this may be illustrated by the following numerical example.

Using Planck's, **h**, the Planck circular frequency, **(V)** has the magnitude, **7.399825*10^{42} Hz**.

For **301** addresses, $\Delta_{(3)}=$ **5.185 $*$ 10^{-3} Hz**, whilst, $\Delta_{(5)}=$ **3.633 $*$ 10^{-48} Hz**.

The last number substantiates the statement made earlier regarding the size of, Δ. Indeed, it could be argued that this far exceeds any reasonable requirement for the magnitude of a finite-difference; The time associated with calculations involving such a size of increment would , to say the least be excessively long, if not completely impractical.

Inspection of Figs2&4 shows that, in each case the magnitude of the quantisation of frequency is made on the basis of a small version of the Fidler diagram.

<u>**The Fidler diagram as the two-dimensional projection of a three-dimensional surface**</u>

A general three-dimensional surface may be represented by the expression: $\zeta = \zeta\ [\ y,x\ ]$, where the notation $\zeta[\quad]$ means a function of **y** and **x** peculiar to ζ, [7].

The structure equation, [3] may be rearranged as: $S_r = S_r\left[\,{^E}/_{(E)}\,,\,{^{(L)}}/_L\,\right]$.

In this instance the form of the above is known explicitly and may be depicted using a right-handed triad of axes where, as shown in Fig5, ${^E}/_{(E)}$ and ${^{(L)}}/_L$ are located perpendicular to each other in a horizontal plane to which the S_r axis is perpendicular.

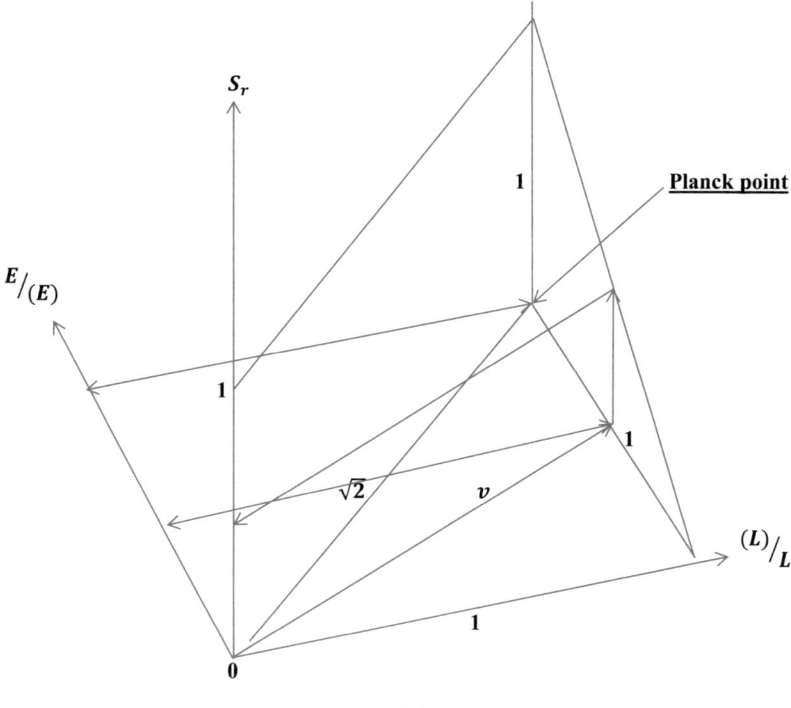

Fig5

In the above diagram we have, for the purpose of clarity, only shown two planes emanating from the S_r axis.

The structure equation,[3] is: $E/_{(E)} = S_r \, {}^{(L)}/_l.$ If we now set ${}^{(L)}/_l$ equal to unity then

we may construct a second Fidler diagram along the line, ${}^{(L)}/_l = 1$, and in a plane parallel to

the $E/_{(E)}$, S_r plane. A vector, **v**, lying in the horizontal plane is drawn from **0**, meeting the

${}^{(L)}/_l$ axis, as shown. Since ${}^{(L)}/_l = 1$, then $E/_{(E)} = S_r$, hence, at the point of intersection

of **v** with the ${}^{(L)}/_l$ axis a vertical line projected upwards from this point will meet the

hypotenuse of the second diagram at a distance, S_r above the point of intersection. If we now draw a line parallel to vector **v** back to intersect the S_r axis, then this forms the plane shown. The upper boundary of this plane is a line of constant S_r; indeed, the plane consists of an infinite number of the constant vector, S_r We may repeat all of the above procedure as many times as desired.

It is then seen that the Fidler diagram is the two-dimensional projection of a helical surface formed from the upper boundaries of an infinite number of planes which wind around the S_r axis.

The extended Fidler diagram

The term 'extended' is somewhat of a misnomer for, in the following, we show that the simplicity of the original diagram may be retained whilst the amount of information that can be extracted may be increased substantially.

It is worth emphasising that a spectrum of electromagnetic radiation relates to the photon and we now proceed to examine other quantities associated with the photon, using the wavelength thereof as a basis.

As noted earlier, the energy, E, associated with a photon is given by: $E = h\nu$. Now, $\nu = c/l$, and so we may write, $E = hc/l$ ------------------- (6).

The spectroscopic wave number, z is defined as, $z = 1/l$. This cannot represent a number for it has dimensions of length^-1, but we adhere to convention and use the title.

We may then write equation (6) as : $E = hcz$ ------------------------ (7). Hence, $E \, \alpha \, z$, for h and c are constants.

It then follows that : $E/_{(E)} = \nu/_{(V)} = z/_{(Z)} = S_r \, {}^{(L)}/_l$ ------------------- (7).

From this it is seen that the ordinate of the Fidler diagram could be replaced by $z/_{(Z)}$.

Whilst this may be of some interest it is not the principal objective of this section.

For a photon we now proceed to define the following quantities:

Specific energy, $s = E/_l$, specific energy intensity, $i = E/_{l^2}$, specific energy density, $d = E/_{l^3}$. Further, we define the frequency ratio, $\nu/_{(V)}$ as e_ν.

Now, $s/_{(S)} = e_s = E/_{(E)} \, {}^{(L)}/_l = S_r \left[{}^{(L)}/_l \right]^2 = S_r \, {}^{(L)}/_l \, {}^{(L)}/_l = \nu/_{(V)} \, {}^{(L)}/_l$ -----(8).

From this we extract the result: $e_s = \nu/_{(V)} \, {}^{(L)}/_l = e_\nu \, {}^{(L)}/_l$ ------------------- (10).

In a similar manner: $i/_{(I)} = e_i = E/_{(E)} \left[{}^{(L)}/_l \right]^2 = S_r \left[{}^{(L)}/_l \right]^3 = e_s \, {}^{(L)}/_l$ --------- (11),

and we may write, $d/_{(D)} = e_d = E/_{(E)} \left[{}^{(L)}/_l \right]^3 = S_r \left[{}^{(L)}/_l \right]^4 = e_i \, {}^{(L)}/_l$ ---------- (12).

Hence, it follows that $\quad {}^{e_s}/_{e_v} = {}^{e_i}/_{e_s} = {}^{e_d}/_{e_i} = {}^{(L)}/_l$ ----------------------- (13).

With the above relationships there is no need to plot parabolic, cubic and quartic curves on the Fidler diagram thus detracting from its simplicity, for all that is needed to determine the specific energy, specific intensity and specific density ratios is already contained in the diagram. Any point on the diagram has coordinates $(e_v, {}^{(L)}/_l)$, and this information, in conjunction with equations (13) is sufficient to determine any, or all of the dependent variables therein.

Using Planck's 'h', the relevant Planck quantities are:

Planck length, $(L) = 4.05134(48) * 10^{-35}\ m.$

Planck circular frequency, $(V) = 4.649445 * 10^{43}\ Hz.$

Planck spectroscopic wave number, $(Z) = 2.468316 * 10^{34}\ m^{-1}.$

Planck specific energy, $(S) = {}^{h(V)}/_{(L)} = 7.604985 * 10^{48}\ J/m\ .$

Planck specific energy intensity, $(I) = 1.877151 * 10^{83}\ J/m^2.$

Planck specific energy density, $(D) = 4.633401 * 10^{118}\ J/m^3\ .$

W M Fidler, March 2019.

References

[1] SI and Metrication Conversion Tables

G Socrates & L J Sapper

Newnes-Butterworths, 1969.

[2] The Lightness of Being

F Wilczek

Penguin books

[3] The Fidler Diagram: A compact and dimensionless representation of the spectrum of

electromagnetic radiation in all media of constant index of refraction

W M Fidler, Jan. 2019.

GRIN Verlag publishers

Catalogue No. v452282, ISBN (book) 9783668876378.

[4] Hua et al

Nature, vol 397, p594.

[5] Hua et al

Physical Review Letters, vol 103, p233602.

[6] Schmidt et al

Nature Photonics, vol4, p776.

[7] Lecture notes on Mathematics

The Graduate School of Thermodynamics & Related Studies.

University of Birmingham, UK.